YOUR KNOWLEDGE HAS VALUE

Bibliographic information published by the German National Library:

The German National Library lists this publication in the National Bibliography; detailed bibliographic data are available on the Internet at http://dnb.dnb.de .

Imprint:

Copyright © 2020 GRIN Verlag
Print and binding: Books on Demand GmbH, Norderstedt Germany
ISBN: 9783346190109

This book at GRIN:

https://www.grin.com/document/538132

Washington Mutwiri

Chemical Reactivity and Bioactivity Rates Of Marine Peptides Hemiasterlin Derivatives. Cancer Treatment Through Conceptual Density Functional Theory

GRIN Verlag

GRIN - Your knowledge has value

Since its foundation in 1998, GRIN has specialized in publishing academic texts by students, college teachers and other academics as e-book and printed book. The website www.grin.com is an ideal platform for presenting term papers, final papers, scientific essays, dissertations and specialist books.

Visit us on the internet:

http://www.grin.com/

http://www.facebook.com/grincom

http://www.twitter.com/grin_com

Study of the Chemical Reactivity and Bioactivity Rates of Marine Peptides Hemiasterlin Derivatives used in the Cancer Treatment through Conceptual Density Functional Theory

Inhaltsverzeichnis

Abstract

This study involved the assessment of the MNI2SX/Def2TZVP and the MNI2SX models to enhance the understanding of the structural composition of marine peptide Hemisterline derivatives A and B used in cancer treatment. The Conceptual Density Functional theory was used in the calculation of molecular properties of the system chemical descriptors during the study. Integration of the active molecular regions into their respective Fukui functions was used in the selection of radical, electrophilic, and nucleophilic attacks. Additionally, the proposed correlation between global hardness and the pKa was used as the basis of deriving accurate predictions for the pKa values while a homology technique was used in the prediction of bioactivity and bioavailability scores of the peptides under investigation.

Introduction

The structural diversity of numerous biologically active metabolites that are found in the marine ecosystems have been used in the development of new categories of agents that can be used in anticancer therapies. The successful development of the anticancer agents has overcome the challenges experienced in the development of drugs from natural resources due to the structural complexity of the agent sourcing process. Despite the challenges several anticancer drugs derived from the marine life agents have been tested and approved as highly effective therapeutic interventions within the past few years. Research reveals that marine life forms contain diverse clinical and preclinical compounds that are potentially vital in the development of new drug formulas for the treatment of human health complications. Researchers have carried out numerous studies to understand the structural and biosynthetic assembling of the marine agents through re-engineering techniques, interdisciplinary development processes, and innovative manipulation within the gene clusters of these agents. These processes are key in enhancing the pharmaceutical properties of the marine agents when compared to the utilization of the natural products directly in the development of human medicine [1, 2].

A wide range of marine species contains bioactive products known as peptides, which contain high amounts of neutraceutical and medicinal agents based on their diverse bioactivities. Pharmacists and medical scientists have leveraged the antimicrobial, neuroprotective, antiviral, immunomodulatory, antioxidative, antidiabetic, analgesic, antiatherosclerotic, cardioprotective, and anxiolytic properties to create drugs that have been used as effective treatments for human diseases. The chemical derivatives of some marine peptides are known to have high demand and commercial value in the pharmaceutical industry due to their important roles in improving patient outcomes in various clinical and preclinical stages of disease treatment. A linear tripeptide known as Hemiasterlin is composed of unique amino acids and cytotoxic properties that are vital in the treatment of leukemia. These properties of Hemiasterlin enhance the clinical treatment of leukaemia by inhibiting the formation of mitotic spindle thus inducing apoptosis and mitotic arrest, which results in tubulin depolymerization [3–8].

The oceanic environment provides habitat to many organisms which are important agents in the manufactured medicine. According to clinical trials that have been carried out to develop medications for cancer, a wide range of marine peptides have been found to have important anti-cancer properties that inhibit growth or kill cancer cells through activities that inhibit different angiogenesis process as well as the tubulin-microtubule balance. The advantage of marine peptides as anticancer agents over the traditional chemotherapeutic interventions is that they do not have extreme side effects on the immune system. Therefore, the use of marine peptides in the development of anticancer peptides is the ideal solution to chemotherapy side effects such as multi-drug resistance, which are common in the use of traditional treatment methods [9]. A wide

3

range of naturally occurring molecules focuses on microtubules as the key drug targets in the treatment of cancer. Combining the marine peptides with the terrestrial anticancer agents such as vinca alkaloids and taxes forms an effective clinical agent that produce tubulin-binding molecules to inhibit the growth of cancer cells [10].

The isolation of cytotoxic peptides from marine sponges results in the formation of hemiasterlins A and B, which are composed of a wide range of unique amino acids such as trimethyltryptophan, N-methyl homovinylogous valine, and tert-leucine. These amino acids contain antimitotic properties which are used in the treatment of different types of cancer [11–13]. This study seeks to assess the chemical properties such as the reactivity and hyperactivity of both hemiasterlin A and B through the application of the density functional theory concept. Understanding the reactivity properties of the hemiasterlin derivatives is important in the use of Fukui functions to represent the peptides reactivity with the molecular systems in the process of developing new drugs. Bioactivity score of the derivatives will also be compared with the descriptors of the conceptual functional density theory [14, 15].

Computational Methodology

The generation of 3D structures and the proposition of their respective low energy conformers in the prediction and calculation of the properties of hemiasterlins in this study was carried out using ChemAxon Calculator Plugins. In the process of geometry optimization, the hemiasterlins with the lowest energy conformation were used while the DFTBA program was used in the optimization of the rests of the conformers. The MNI12SX/Def2TZVP/H20 model was used in the re-optimization of the five conformers having the lowest energy. Consequently, the real minimal approach was used in the confirmation of the optimized hemiasterlin structures through the application of the vibrational frequency analysis technique. In the process calculating

4

the electronic properties for chemical reactivity of the derivatives, A and B involved the use of MN12SX/Def2TZVP/H2O model through the optimized molecular structures.

Results and discussion

ChemAxon Calculator Plugins were used in the process of deriving the molecular structures and the bioactivity properties of the conformers. The optimization and re-optimization of the conformers was carried out using the DFTBA program and the MNISX/DefTZVP/H2O model respectively as explained in the Computational Methodology section [16]. The graphical sketches of the molecular structures of the hemiasterlin, Hemiasterlin A, and Hemiasterlin B are shown in figure 1 below. The Density Functional Tight-Binding method was used in the re-optimization of the molecular structures while the MNISX density functional method combined with the SMD solvent model, and the DefTZVP were used in the second optimization of the molecular structures. The MNISX/DefTZVP/H2O model was used in determining the electronic properties of each molecular structure after using calculation analysis procedures to determine whether all the molecular structures correspond to their respective minimum energy requirements. According to Becke, a common misconception exists in the integration of the electronic ground states and the adiabatic connection since the superiority of the KS model is not recognized due to its minimum molecular energy [17]. Baerends et al. state that the level of energy excitation within a KS system is used as an effective measure of the optimization to the molecular optical gap [18]. HOMO-LUMO gap of the KS model is used to approximate the citation energy within the KS model based which is a basic requirement in determining the consistency with the molecular structures [19]. Ground state calculations are used in determining the optimal wavelength that is infused in the marine peptides that contain the Hemiasterlins anticancer properties based on the pre-determined density function to find the respective λ_{max} values through the application of theoretical models to establish the HOMO-LUMO gaps.

Table 1

This table shows the electronic energies of different molecular systems containing the neutral derivatives A and B of Hemiasterlin, the optimal absorption wavelength λ_{max}, the eV HOMO-LUMO gaps, and the orbital energies of HOMO and LUMO, which are calculated using the SMD parametrization solvent model, the Def2TZVP, and the MNI2SX density functional mode.

Molecule	Total Electronic Energy	HOMO	LUMO	HOMO-LUMO Gap	λ_{max}
Hemiasterlin	-1690.078	-5.542	-1.907	3.545	350
Hemiasterlin A	-1650.793	-5.542	-1.907	3.635	341
Hemiasterlin B	-1611.505	-5.471	-1.927	3.544	350

3.1 Computing Global Reactivity Descriptors

According to Frau et al., the evaluation of marine peptides and melanoidins in the generation of HOMO and LUMO energies that are required in the verification of the levels of agreement with the estimated Koopman's theorem based on the combination of MN12SX and Def2TZVP [20–27]. This justifies the application of the proposed KID technique.

6

Considering the KID technique used on the previous studies being integrated into the finite difference approximation [20–27], the following expressions can be used to define the global reactivity descriptors.

Electronegativity -> $x = -\frac{1}{2}(1 + A) \approx \frac{1}{2}(\epsilon L + \epsilon H)$ [28, 29]

Global Hardness -> $\eta = (I - A) \approx (\epsilon L - \epsilon H)$ [28, 29]

Electrophilicity -> $\omega = \frac{\mu 2}{2\eta} = \frac{(I+A)^2}{4(I-A)} \approx \frac{(\epsilon L + \epsilon H)^2}{4(\epsilon L - \epsilon H)}$ [30]

Electrodonating Power -> $\omega^- = \frac{(3I+A)^2}{16(I-A)} \approx \frac{(3\epsilon H + \epsilon L)^2}{16\eta}$ [31]

Electroaccepting Power -> $\omega^- = \frac{(I+3A)^2}{16(I-A)} \approx \frac{(\epsilon H + 3\epsilon L)^2}{16\eta}$ [31]

Net Electrophilicity -> $\Delta\omega^\pm = \omega^+ - (-\omega^-) = \omega^+ + \omega^-$ [32]

Given that

ϵ_H represents the HOMO energies while ϵ_L represents LUMO energies.

Table 2: The MN12SX density function is used in the computation of the HOMO and LUMO energy values corresponding to the global reactivity descriptors in Hemiasterlin and its derivatives A and B

Molecule	Electronegativity	Global Hardness	Electrophilicity
Hemiasterlin	3.679	3.545	1.909
Hemiasterlin A	3.724	3.635	1.908
Hemiasterlin B	3.669	3.544	1.931

7

Molecule	Electrodonating Power	Electroaccepting Power	Net Electrophilicity
Hemiasterlin	5.880	2.200	8.080
Hemiasterlin A	5.906	2.181	8.087
Hemiasterlin B	5.932	2.233	8.165

According to the table values, the electrodonating property of the molecular structure of peptides is more vital compared to its electroaccepting properties. However, there is no significant variation in the values of the global descriptors of the molecular structures of the respective peptides.

3.2 Computation of the Local Reactivity Descriptors

The expression of local reactivity descriptors is shown below;

Nucleophilic Fukui Function -> $f^+ (r) = \rho N + 1^{(r)} - \rho N$ [28, 29]

Electrophilic Fukui Function -> $f (r) = \rho N (r) - \rho N - 1(r)$ [28, 29]

Dual Descriptor -> $\Delta f(r) = (\frac{\partial\, f(r)}{\partial\, N})_{v(r)}$ [33–35] [36, 37]

Given point r, $\rho N+1(r)$ represents the electronic density of the system $N + 1$, $\rho N (r)$ represents the electronic density system N, and $\rho N - 1(r)$ represents the electronic density system $N - 1$.

The figure 2 below shows the electrophilic and the Nucleophilic Fukui functions $f^-(r)$ and $f^+(r)$ for the respective Hemiasterlins.

3.3 Computation of the Marine pKas Anticancer Peptides

The previous discussion focused on the application of the conceptual DFT descriptors to evaluate the computation prediction of the pKas peptides where it was established that the pKa = $16.3088 - 0.868\eta$ relationship would play an important role in the initial prediction of complex peptides, which are important in the manufacture of medical drugs [38]. Given the biological level of PH, the peptides under study exist as neutral molecules and are still considered to be neutral during the pKa computations [38]. The pKa relationship is also important in the optimization of the molecular structure of every conformer as well as the computation of the pKa values for all molecules given the η values shown in table 2. The computational results of the pKa values for the hemiasterlin molecules are shown in Table 3 below.

Table 3: The pKa value representation of Hemiasterlin with its derivatives A and B

Molecule	pKa value
Hemiasterlin	13.38
Hemiasterlin A	13.30
Hemiasterlin B	13.38

The pKa values shown in table 3 indicate that the computational methodology used is effective in the differentiation of the respective pKa values for all peptides molecules irrespective of the significance of the difference. The pKa values of these peptides are important in the manufacture of pharmaceutical drugs by explaining the procedures used in drug delivery and their respective action mechanisms.

Bioactivity Scores

According to Leeson et al., it's important to check the species level of compliance of a potential therapeutic drug to the Lipinski rule of five, which explains whether the compound contains drug certain drug properties [39]. Molinsipratoon and MolSoft software used to compute the molecular drug properties in a compound as shown in Table 4 given that miLog*P* is a representation of the water partition coefficient. The rate of violations of the Lipinski Rule of Five is measured using nviol while TPSA represents the polar surface area of the molecule. The hydrogen bond donors and the hydrogen bond acceptors are represented by nOHNH and the nON respectively.

Table 4 below shows the application of the Lipinski Rule of Five in the calculation the molecular properties of anticancer marine peptides of the Hemiasterlin family. Volume represents the molecular volume of the peptides while MW represents the molecular weight of the peptides.

Molecule	miLogP	TPSA	nAtoms	Non	NOHNH
Hemiasterlin	4.65	103.67	38	8	3
Hemiasterlin A	4.48	114.52	37	8	4
Hemiasterlin B	4.00	114.52	36	8	4
Molecule	Nviol	Nrotb	Volume	MW	
Hemiasterlin	1	11	525.87	526.72	

| Hemiasterlin A | 1 | 11 | 508.92 | 512.70 | |
| Hemiasterlin B | 0 | 11 | 492.69 | 498.67 | |

The degree of oral bioavailability of the marine molecules that can be potentially used in the manufacture of drugs is measured using the Lipinski Rule of Five by determining the molecules that possess drug-like properties. However, this technique cannot be applied in measuring the bioavailability of the peptides due to the existence of hydrogen bonds and molecular weight properties as shown in Table 4 above.

This study applied a different technique in the evaluation of the chemical structure of other compounds that were predicted to possess similar pharmacological properties as the Hemiasterlin and other compounds under study. As illustrated in the computational methodology, the evaluation of pharmacological properties of different compounds in the process of determining the bioactivity scores can be carried out Using Molinspiration Software based on the variability of the drug targets as shown in table 5 below. According to the table, the organic molecules whose bioactivity score is less than zero are considered to be active while the organic molecules whose bioactivity score are between zero and negative five are considered to be moderately active and the organic molecules with a score of less than negative five are considered to be inactive. All peptides that were studied during this study were found to have moderate bioactivity scores with respect to enzymatic reactivity.

Conclusion

This paper describes a study carried out to investigate the reactivity properties of

Hemiasterlin and its derivatives using the density functional theory to explain how the molecular

interactions of these marine peptides can be used in anticancer drugs.

Table 5 below shows the bioactivity scores of hemiasterlin and its derivatives based on

the interactions with various enzyme inhibitors, GPCR Ligand, the protease inhibitors, the ion

channel modulators, the kinase inhibitors, and the nuclear receptors.

Molecule	GPCR Ligand	Ion Channel Modulator	Kinase Inhibitor
Hemiasterlin	0.51	0.02	0.01
Hemiasterlin A	0.56	0.13	0.04
Hemiasterlin B	0.49	0.12	0.06
Molecule	Nuclear Receptor Lingad	Protease Inhibitor	Enzyme Inhibitor
Hemiasterlin	0.19	0.62	0.43
Hemiasterlin A	0.18	0.68	0.45
Hemiasterlin B	0.13	0.58	0.35

The development of new pharmaceutical drugs, especially in cancer treatment requires

extensive knowledge of the molecular bioactivity scores of different global and local peptides.

Similarly, the values of the chemical hardness of some therapeutic peptides have been used as

the basis of determining their respective pKa values as proposed in the computational

methodology. This information that is obtained enhances the understanding of properties such as the chemical reactivity and the water solubility of the peptides. Moreover, prediction of the molecular properties of the peptides using various methodologies as described in the literature can also be used in the computation of the bioavailability values. As a result, the bioactivity levels can be quantified on the basis of their respective descriptors in the characterization process of the peptide bioactivity with the GPCR Ligand and the protease inhibitors.

References

[1] W. H. Gerwick, B. S. Moore, Lessons from the Past and Charting the Future of Marine Natural Products Drug Discovery and Chemical Biology, Chemistry & Biology 19 (1) (2012) 85–98.

[2] A. Joseph, Investigating Seafloors and Oceans: From Mud Volcanoes to Giant Squid, Elsevier, Amsterdam, Netherlands, 2017.

[3] R. Cheung, T. Ng, J. Wong, Marine Peptides: Bioactivities and Applications, Marine Drugs 13 (7) (2015) 4006–4043.

[4] M. Mehbub, J. Lei, C. Franco, W. Zhang, Marine Sponge Derived Natural Products between 2001 and 2010: Trends and Opportunities for Discovery of Bioactives, Marine Drugs 12 (8) (2014) 4539–4577.

[5] G. M. Cragg, D. J. Newman, Natural Products: A Continuing Source of Novel Drug Leads, Biochimica et Biophysica Acta (BBA) - General Subjects 1830 (6) (2013) 3670–3695.

[6] R. Pallela, H. Ehrlich (Eds.), Marine Sponges: Chemicobiological and Biomedical Applications, Springer, India, 2016.

[7] R. Sable, P. Parajuli, S. Jois, Peptides, Peptidomimetics, and Polypeptides from Marine Sources: A Wealth of Natural Sources for Pharmaceutical Applications, Marine Drugs 15 (4) (2017) 124.

[8] S. Agrawal, A. Adholeya, S. K. Deshmukh, The Pharmacological Potential of Non-ribosomal Peptides from Marine Sponge and Tunicates, Frontiers in Pharmacology 7 (2016) 333.

[9] H. K. Kang, M.-C. Choi, C. H. Seo, Y. Park, Therapeutic Properties and Biological Benefits of Marine-Derived Anticancer Peptides, International Journal of Molecular Sciences 19 (3) (2018) 919. 12

[10] M. J. Mart'ın, L. Coello, R. Fern'andez, F. Reyes, A. Rodr'ıguez, C. Murcia, M. Garranzo, C. Mateo, F. S'anchez-Sancho, S. Bueno, C. de Eguilior, A. Francesch, S. Munt, C. Cuevas, Isolation and First Total Synthesis of PM050489 and PM060184, Two New Marine Anticancer Compounds, Journal of the American Chemical Society 135 (27) (2013) 10164–10171.

[11] R. Pangestuti, S.-K. Kim, Bioactive Peptides of Marine Origin for the Prevention and Treatment of Non-Communicable Diseases, Marine Drugs 15 (3) (2017) 67.

[12] D. Giordano, M. Costantini, D. Coppola, C. Lauritano, L. N. nez Pons, N. Ruocco, G. di Prisco, A. Ianora, C. Verde, Chapter five - biotechnological applications of bioactive peptides from marine sources, Vol. 73 of Advances in Microbial Physiology, Academic Press, 2018, pp. 171–220.

[13] S.-K. Kim, Handbook of Anticancer Drugs from Marine Origin, Springer International Publishing Imprint Springer, Cham, 2015.

[14] G. K. Gupta, V. Kumar, Chemical Drug Design, Walter de Gruyter GmbH, Berlin, 2016.

[15] M. Gore, U. B. Jagtap, Computational Drug Discovery and Design, Springer Science+Business Media, LLC, New York, 2018.

[16] M. J. Frisch, G. W. Trucks, H. B. Schlegel, G. E. Scuseria, M. A. Robb, J. R. Cheeseman, G. Scalmani, V. Barone, B. Mennucci, G. A. Petersson, H. Nakatsuji, M. Caricato, X. Li, H. P. Hratchian, A. F. Izmaylov, J. Bloino, G. Zheng, J. L. Sonnenberg, M. Hada, M. Ehara, K. Toyota, R. Fukuda, J. Hasegawa, M. Ishida, T. Nakajima, Y. Honda, O. Kitao, H. Nakai, T. Vreven, J. A. Montgomery, Jr., J. E. Peralta, F. Ogliaro, M. Bearpark, J. J. Heyd, E. Brothers, K. N. Kudin, V. N. Staroverov, R. Kobayashi, J. Normand, K. Raghavachari, A. Rendell, J. C. Burant, S. S. Iyengar, J. Tomasi, M. Cossi, N. Rega, J. M. Millam, M. Klene, J. E. Knox, J. B. Cross, V. Bakken, C. Adamo, J. Jaramillo, R. Gomperts, R. E. Stratmann, O. Yazyev, A. J. Austin, R. Cammi, C. Pomelli, J. W. Ochterski, R. L. Martin, K. Morokuma, V. G. Zakrzewski, G. A. Voth, P. Salvador, J. J. Dannenberg, S. Dapprich, A. D. Daniels, O. Farkas, J. B. Foresman, J. V. Ortiz, J. Cioslowski, D. J. Fox, Gaussian 09 Revision E.01, Gaussian Inc., Wallingford CT, 2016 (2016). 13

[17] A. D. Becke, Vertical Excitation Energies From the Adiabatic Connection, The Journal of Chemical Physics 145 (19) (2016) 194107.

[18] E. J. Baerends, O. V. Gritsenko, R. van Meer, The Kohn-Sham Gap, the Fundamental Gap and the Optical Gap: The Physical Meaning of Occupied and Virtual Kohn-Sham Orbital Energies, Physical Chemistry Chemical Physics 15 (39) (2013) 16408–16425.

15

[19] R. van Meer, O. V. Gritsenko, E. J. Baerends, Physical Meaning of Virtual Kohn-Sham Orbitals and Orbital Energies: An Ideal Basis for the Description of Molecular Excitations, Journal of Chemical Theory and Computation 10 (10) (2014) 4432–4441.

[20] J. Frau, D. Glossman-Mitnik, Molecular Reactivity and Absorption Properties of Melanoidin Blue-G1 through Conceptual DFT, Molecules 23 (3) (2018) 559–15.

[21] J. Frau, D. Glossman-Mitnik, Conceptual DFT Study of the Local Chemical Reactivity of the Dilysyldipyrrolones A and B Intermediate Melanoidins, Theoretical Chemistry Accounts 137 (5) (2018) 1210.

[22] J. Frau, D. Glossman-Mitnik, Conceptual DFT Study of the Local Chemical Reactivity of the Colored BISARG Melanoidin and Its Protonated Derivative, Frontiers in Chemistry 6 (136) (2018) 1–9.

[23] J. Frau, D. Glossman-Mitnik, Molecular Reactivity of some Maillard Reaction Products Studied through Conceptual DFT, ContemporaryChemistry 1 (1) (2018) 1–14.

[24] J. Frau, D. Glossman-Mitnik, Computational Study of the Chemical Reactivity of the Blue-M1 Intermediate Melanoidin, Computational and Theoretical Chemistry 1134 (2018) 22–29.

[25] J. Frau, D. Glossman-Mitnik, Chemical Reactivity Theory Applied to the Calculation of the Local Reactivity Descriptors of a Colored Maillard Reaction Product, Chemical Science International Journal 22 (4) (2018) 1–14.

[26] J. Frau, D. Glossman-Mitnik, Blue M2: An Intermediate Melanoidin Studied via Conceptual DFT, Journal of Molecular Modeling 24 (138) (2018) 1–13. 14

16

[27] J. Frau, N. Flores-Holguín, D. Glossman-Mitnik, Chemical Reactivity Properties, pKa Values, AGEs Inhibitor Abilities and Bioactivity Scores of the Mirabamides A–H Peptides of Marine Origin Studied by Means of Conceptual DFT, Marine Drugs 16 (9) (2018) 302–19.

[28] R. Parr, W. Yang, Density-Functional Theory of Atoms and Molecules, Oxford University Press, New York, 1989.

[29] P. Geerlings, F. De Proft, W. Langenaeker, Conceptual Density Functional Theory, Chemical Reviews 103 (2003) 1793–1873.

[30] R. Parr, L. Szentpaly, S. Liu, Electrophilicity Index, Journal of the American Chemical Societ 121 (1999) 1922–1924.

[31] J. Gázquez, A. Cedillo, A. Vela, Electrodonating and Electroaccepting Powers, Journal of Physical Chemistry A 111 (10) (2007) 1966–1970.

[32] P. Chattaraj, A. Chakraborty, S. Giri, Net Electrophilicity, Journal of Physical Chemistry A 113 (37) (2009) 10068–10074.

[33] C. Morell, A. Grand, A. Toro-Labbé, New Dual Descriptor for Chemical Reactivity, Journal of Physical Chemistry A 109 (2005) 205–212.

[34] C. Morell, A. Grand, A. Toro-Labbé, Theoretical Support for Using the $\Delta f(r)$ Descriptor, Chemical Physics Letters 425 (2006) 342–346.

[35] J. I. Martínez-Araya, Revisiting Caffeate's Capabilities as a Complexation Agent to Silver Cation in Mining Processes by means of the Dual Descriptor – A Conceptual DFT Approach, Journal of Molecular Modeling 18 (2012) 4299–4307.

17

[36] J. I. Mart´ınez-Araya, Explaining Reaction Mechanisms Using the Dual Descriptor: A Complementary Tool to the Molecular Electrostatic Potential, Journal of Molecular Modeling 19 (7) (2012) 2715–2722. [37] J. I. Mart´ınez-Araya, Why is the Dual Descriptor a More Accurate Local Reactivity Descriptor than Fukui Functions?, Journal of Mathematical Chemistry 53 (2) (2015) 451–465.

[38] J. Frau, N. Hern´andez-Haro, D. Glossman-Mitnik, Computational Prediction of the pKas of Small Peptides through Conceptual DFT Descriptors, Chemical Physics Letters 671 (2017) 138–141. 15

[39] P. Leeson, Drug Discovery: Chemical Beauty Contest, Nature 481 (7382) (2012) 455–456.